Math Facts Fun

Fun with numbers for all ages and abilities

Developed by
Catherine McGrew Jaime
and
Cheryl McGrew Holle

Creative Learning Connection
8006 Old Madison Pike
Madison, AL 35758
U.S.A.

More fun games available at
www.CreativeLearningConnection.com

Math Facts Fun

Fun with numbers for all ages and abilities

Developed by Catherine McGrew Jaime and Cheryl McGrew Holle

www.CatherineJaime.com

Object of Game:

To be the first to cover 3, 4, or 5 numbers in a row (depending on the level).

Variations:

You have different levels, boards, and suggestions, to get you started, but feel free to modify the rules as you play the game, according to what works best for your group's preferences and abilities.

Level 1
One Six-Sided Die
Goal = 3 in a Row

The first level of math is <u>Counting/Number Recognition</u>. The first game board shows the numbers 1 – 6 in random order. Much younger kids can match the number of dots on the die they roll with the numbers on their board.

Playing the Game:
Each player chooses a color of markers.
Play begins with the youngest player and rotates to the left.
On their turn each player rolls a die and chooses a number to cover on the board. If that number is unavailable they re-roll the die.

Winning the Game:
The first player to successfully cover 3 numbers in a row (on the board, not numerically) wins the game.

Level 1
Counting/Number Recognition
One Six-Sided Die

4	1	2
1	6	5
5	2	2
3	1	6
5	4	3
2	3	6
3	1	4

Level 2
Two Six-Sided Dice
Goal = 3 in a Row

Pre-addition is counting two sets of things together as one. Here we have provided an opportunity for kids to practice that with two dice. This side of the game board contains the numbers 2 – 12, in order to practice this important pre-addition skill.

Playing the Game:
Each player chooses a color of markers.
Play begins with the youngest player and rotates to the left.
On their turn each player rolls the dice and chooses a number to cover on the board. If that number is unavailable they re-roll the die.

Winning the Game:
As before, the first player to successfully cover 3 numbers in a row wins the game.

Level 2

Pre-Addition
Two Six-Sided Dice

4	10	2
7	8	5
5	7	8
9	11	6
12	4	3
2	11	6
3	10	9

Level 3
Two Six-Sided Dice
Goal = 3 in a Row

Once players are doing some basic <u>addition</u> and <u>subtraction</u>, they move to the second game board – with numbers **0** -12. With 2 dice, each of these numbers can be obtained by addition and/or subtraction.

With most rolls, there will be two choices to pick from, for example:
>   With the roll of a 1 and a 2, they can cover 3 or 1
>   With the roll of a 2 and a 6, they can cover 4 or 8

Playing the Game:
Each player chooses a color of markers.
Play begins with the youngest player and rotates to the left.
On their turn each player rolls the dice and chooses a number to cover on the board. If that number is unavailable they re-roll the die.

Winning the Game:
And again, as before, the first player to successfully cover 3 numbers in a row wins the game.

Level 3

Addition and Subtraction
Two Six-Sided Dice

4	10	2
7	8	5
5	7	8
9	1	6
12	4	3
2	11	6
3	1	0

Level 3a
Two Twelve-Sided Dice
Goal = 3 in a Row

Once players are doing some basic <u>addition</u> and <u>subtraction</u>, they move to the second game board – with numbers **0** -12. With 2 dice, each of these numbers can be obtained by addition and/or subtraction.

With most rolls, there will be two choices to pick from, for example:

      With the roll of a 1 and a 2, they can cover 3 or 1

      With the roll of a 2 and a 6, they can cover 4 or 8

Playing the Game:

Each player chooses a color of markers.

Play begins with the youngest player and rotates to the left.

On their turn each player rolls the dice and chooses a number to cover on the board. If that number is unavailable they re-roll the die.

Winning the Game:

And again, as before, the first player to successfully cover 3 numbers in a row wins the game.

Level 3a
Addition & Subtraction
Two Twelve-Sided Dice

3	15	12
16	5	2
9	21	7
8	1	11
20	6	24
18	22	10
4	13	19
23	17	14

Level 4
Two Six-Sided Dice
Goal = 3 in a Row

It doesn't take long before players want to start multiplying and dividing, also. With 2 dice, most numbers up to 36 can be calculated with <u>addition</u>, <u>subtraction</u>, <u>multiplication</u>, and/or <u>division</u>, but not all. So we've designed this special game board to include all the numbers that can be made with 2 dice.

Now the options for each turn increase dramatically. For example:
    With the roll of a 2 and a 6,
        they can cover
            3 (6/2)   8 (6+2)
            4 (6-2)   or 12 (6*2)

Playing the Game:
Each player chooses a color of markers.
Play begins with the youngest player and rotates to the left.
On their turn each player rolls the dice and chooses a number to cover on the board. If that number is unavailable they re-roll the die.

Winning the Game:
And, guess what? The first player to successfully cover 3 numbers in a row wins the game! (Starting to see a pattern here?)

Level 4

Addition, Subtraction, Multiplication, Division
Two Six-Sided Dice

36	1	25
3	15	5
30	18	8
9	16	6
12	4	10
2	11	24
7	20	0

Level 4a
Two Six-Sided Dice
Goal = 3 in a Row

It doesn't take long before players want to start multiplying and dividing, also. With 2 dice, most numbers up to 36 can be calculated with <u>addition</u>, <u>subtraction</u>, <u>multiplication</u>, and/or <u>division</u>, but not all. So we've designed this special game board to include all the numbers that can be made with 2 dice.

Now the options for each turn increase dramatically. For example:
>   With the roll of a 2 and a 6,
>
>   they can cover

3 (6/2)	8 (6+2)
4 (6-2)	or 12 (6*2)

Playing the Game:
Each player chooses a color of markers.
Play begins with the youngest player and rotates to the left.
On their turn each player rolls the dice and chooses a number to cover on the board. If that number is unavailable they re-roll the die.

Winning the Game:
And, guess what? The first player to successfully cover 3 numbers in a row wins the game! (Starting to see a pattern here?)

Level 4a

Addition, Subtraction, Multiplication, Division
Two Twelve-Sided Dice

21	132	84	11	35	28
66	27	144	22	5	77
18	1	110	15	36	19
24	44	50	3	10	88
80	14	0	99	63	24
8	72	42	13	120	7
25	40	6	90	30	45
16	56	70	49	12	36
108	32	64	33	20	54
9	60	96	4	55	2
121	23	81	17	100	48

Level 5
Three Six-Sided Dice
Goal = 4 in a Row

For you older players, you're obviously starting to think beyond 2 numbers and one operation, so we're introducing a 3rd dice. With 3 dice, it becomes much more challenging. With 3 dice, you have to use all 3 numbers, but you can use them in any order, with various operations. For instance:

With the roll of a 2, 3, and 4
options include:

(3+2)-4=1	(3*2)-4=2	(4+2)-3=3	(3-2)+4=5	(4-2)*3=6
2+3+4=9	(3*2)+4=10	(4*2)+3=11	(4*3)+2=14	4*3*2=24

Playing the Game:
Each player chooses a color of markers.
Play begins with the youngest player and rotates to the left.
On their turn each player rolls the dice and chooses a number to cover on the board. If that number is unavailable they re-roll the die.

Winning the Game:
Time for a curve ball! The first player to successfully cover **four** numbers in a row wins the game! (Bet you didn't see that coming, huh?)

Level 5

Addition, Subtraction, Multiplication, Division
Three Six-Sided Dice

11	25	19	8	22	3
40	14	5	26	13	37
7	32	21	42	1	30
38	15	36	12	33	6
18	24	9	28	10	23
27	2	31	17	41	34
16	35	20	39	4	29

Level 6
Four Six-Sided Dice
Goal = 5 in a Row

Now things get really interesting! You have 4 dice to play with. All the numbers up to 130 can be gotten with 1 or more combinations of 4 dice. Adults enjoy this level as much as the older kids. It's a great way to shake loose the mental cobwebs!

Again, any combination of operations can be used, as long as each dice is used one time. Most rolls will give many different choices, so the opportunities to play strategically increase with this level.

One roll will be able to be used in many, many ways at this level.

Example: rolling 1,2,3, and 4

$1+2+3+4=10$

$1*2*3*4=24$

$(4+3)-(2+1)=4$

$(4+3)*(2+1)=21$

$(4-3)+(2+1)=4$

$(4-3)*(2+1)=3$

$(2+1)-(4-3)=2$

$(2+1)/(4-3)=3$

$(1+3+4)-2=6$

$(1+3+4)+2=10$

$(1+3+4)*2=16$

$(1+3+4)/2=4$

$(1*3*4)-2=10$

$(1*3*4)+2=14$

$(1*3*4)*2=24$

$(1*3*4)/2=6$

and on it goes…

Playing the Game:
Each player chooses a color of markers.
Play begins with the youngest player and rotates to the left.
On their turn each player rolls the dice and chooses a number to cover on the board. If that number is unavailable they re-roll the die.

Winning the Game:
The first player to successfully cover **5** in a row wins the game. (But I think you already knew that, didn't you?)

Level 6
Addition, Subtraction, Multiplication, Division
Four Six-Sided Dice

6	42	90	16	104	75	47	4	108	30
77	129	1	88	9	40	91	54	19	117
29	110	67	39	84	64	15	43	121	62
71	25	49	80	23	36	120	66	10	85
114	60	11	37	100	61	2	113	46	130
3	112	74	103	55	28	70	58	21	87
86	17	96	5	72	116	31	95	105	41
115	56	22	93	45	78	122	7	24	126
33	83	109	51	106	12	73	38	111	76
59	119	8	68	32	102	50	79	13	128
107	44	99	53	101	63	18	97	52	81
94	26	89	14	82	48	92	65	27	118
57	124	35	123	69	20	125	34	127	98

Math
Extras

Skip Counting

Skip Counting prepares students for multiplying. The enclosed charts may be used to introduce students to the concept, and to practice with.

Prime Numbers

Prime Numbers are numbers with no factors, other than themselves and one. Prime Numbers will come up many times later in Math. Introducing the concept early often takes away the "fear" associated with them. Understanding prime numbers, as well as knowing how to find them, will help students as they move into Algebra. In this game, knowing what numbers are prime numbers will help players as thy look for combinations to obtain needed numbers. For example, knowing that 13 is a prime number lets a player know that they can't get there by multiplying 2 numbers together. So, if they need a 13, they will have to get there another way: (6*2)+1 or (6*3)-5, for instance.

"Sieve of Eratosthenes"

A 2,000-year-old method of finding Prime Numbers, and a great way to introduce skip counting and prime numbers. Have fun with it! Better yet, let your students have fun with it.

Hundred Chart

A hundred chart can be used for so many things…Skip counting can be practiced on it, the "sieve" can be done on it, patterns in numbers can be discovered…Explore its possibilities, and let your students explore it! Ruth Beechick has a great little book,
An Easy Start in Arithmetic, which gives many more ideas for using a Hundred Chart.

Addition and Multiplication Tables

Addition and Multiplication Tables are nothing new. But sometimes we think of them as "cheating". Encourage, rather than discourage, your students from using them during the game! Every time they look up something on the table, it's being reinforced in their brain, and eventually they will come to need the tables less and less.

Skip Counting

By 2's

2	4	6	8	10	12	14	16
18	20	22	24	26	28	30	32
34	36	38	40	42	46	48	50

By 3's

3	6	9	12	15	18	21	24
27	30	33	36	39	42	45	48
51	54	57	60	63	66	69	72

By 5's

5	10	15	20	25
30	35	40	45	50
55	60	65	70	75
80	85	90	95	100

By 10's

10	20	30	40	50
60	70	80	90	100

Prime Numbers

2	59
3	61
5	67
7	71
11	73
13	79
17	83
19	89
23	97
29	101
31	103
37	107
41	109
43	113
47	127
53	

"The Sieve of Eratosthenes"

or *Skip Counting Meets Prime Numbers...*

Eratosthenes was born in Northern Africa in the third century before Christ. He was a mathematician, astronomer, philosopher and much more. He discovered a method to "sift out" prime numbers from a long list of consecutive whole numbers. That method, "The Sieve of Eratosthenes", can be used to show all the prime numbers up to any number; in our case it was used to find the prime numbers up to 130[1] (as shown on our **Prime Numbers Chart**).

1. The number 1 is universally considered "not prime" – so it is crossed out.[2]
     Looks like this in the chart: ~~1~~

2. The number 2 is the first prime number, and thus is circled (○ in front of it in this chart). Every multiple of 2 is crossed out. (Time to skip count by 2's.) Those numbers are "not prime" – because they each have a factor of 2. (i.e. prime numbers can only be *odd* numbers.)
     Looks like this in the chart: ~~4~~

3. The next number that is not crossed out is the number 3. Therefore, it is the next prime number, and is thus circled. Skip count by 3's and cross out each of those numbers, the numbers with 3 as a factor.
     Looks like this in the chart:

4. The next prime number is the number 5 – not yet circled nor crossed out. It needs to be circled, and then you skip count by 5's – crossing out each of these numbers which have 5 as a factor.
     Looks like this in the chart:

5. This continues until all numbers have been circled or crossed out. All the numbers that have been circled are prime – they have no factors other than themselves and the number 1.
     The next "crossed out numbers will look like this in the chart:
          ^ (skip counting by 7's), < (skip counting by 11's),
          >(skip counting by 13's), +(skip counting by 17's,19s, and the rest....)

~~1~~	○2	○3	~~4~~	○5	~~6~~	○7	~~8~~	~~9~~	~~10~~
○11	~~12~~	○13	^~~14~~	~~15~~	~~16~~	○17	~~18~~	○19	~~20~~
^21	<~~22~~	○23	~~24~~	~~25~~	>~~26~~	~~27~~	^~~28~~	○29	~~30~~
○31	~~32~~	<~~33~~	+~~34~~	^~~35~~	~~36~~	○37	+~~38~~	>~~39~~	~~40~~
○41	^~~42~~	○43	<~~44~~	~~45~~	+~~46~~	○47	~~48~~	^49	~~50~~
+~~51~~	>~~52~~	○53	~~54~~	<~~55~~	^~~56~~	+~~57~~	+~~58~~	○59	~~60~~
○61	~~62~~	^~~63~~	~~64~~	>~~65~~	<~~66~~	○67	+~~68~~	+~~69~~	^~~70~~
○71	~~72~~	○73	+~~74~~	~~75~~	+~~76~~	<^~~77~~	>~~78~~	○79	~~80~~
~~81~~	+~~82~~	○83	^~~84~~	+~~85~~	+~~86~~	+~~87~~	<~~88~~	○89	~~90~~
>^91	+~~92~~	+~~93~~	+~~94~~	+~~95~~	~~96~~	○97	^~~98~~	<~~99~~	~~100~~
○101	+~~102~~	○103	>~~104~~	~~105~~	+~~106~~	○107	~~108~~	○109	<~~110~~
+~~111~~	^~~112~~	○113	+~~114~~	+~~115~~	+~~116~~	>~~117~~	+~~118~~	+^119	~~120~~
<121	+~~122~~	+123	+~~124~~	~~125~~	^~~126~~	○127	~~128~~	+~~129~~	>~~130~~

Feel free to try this with any size number chart you would like!
(We've included a "Hundreds" chart in the packet to get you started.)

---

[1] Prime numbers will end up circled (or in the case of my chart, with a ○ in front of them), others will be crossed out.
[2] No clue why! But lists of prime numbers always start with the number 2, so who am I to argue?

Hundred Chart

1	2	3	4	5	6	7	8	9	10
11	12	13	14	15	16	17	18	19	20
21	22	23	24	25	26	27	28	29	30
31	32	33	34	35	36	37	38	39	40
41	42	43	44	45	46	47	48	49	50
51	52	53	54	55	56	57	58	59	60
61	62	63	64	65	66	67	68	69	70
71	72	73	74	75	76	77	78	79	80
81	82	83	84	85	86	87	88	89	90
91	92	93	94	95	96	97	98	99	100

Addition and Subtraction Table

+/-	1	2	3	4	5	6	7	8	9	10	11	12
1	2	3	4	5	6	7	8	9	10	11	12	13
2	3	4	5	6	7	8	9	10	11	12	13	14
3	4	5	6	7	8	9	10	11	12	13	14	15
4	5	6	7	8	9	10	11	12	13	14	15	16
5	6	7	8	9	10	11	12	13	14	15	16	17
6	7	8	9	10	11	12	13	14	15	16	17	18
7	8	9	10	11	12	13	14	15	16	17	18	19
8	9	10	11	12	13	14	15	16	17	18	19	20
9	10	11	12	13	14	15	16	17	18	19	20	21
10	11	12	13	14	15	16	17	18	19	20	21	22
11	12	13	14	15	16	17	18	19	20	21	22	23
12	13	14	15	16	17	18	19	20	21	22	23	24

Multiplication and Division Table

x /÷	1	2	3	4	5	6	7	8	9	10	11	12
1	1	2	3	4	5	6	7	8	9	10	11	12
2	2	4	6	8	10	12	14	16	18	20	22	24
3	3	6	9	12	15	18	21	24	27	30	33	36
4	4	8	12	16	20	24	28	32	36	40	44	48
5	5	10	15	20	25	30	35	40	45	50	55	60
6	6	12	18	24	30	36	42	48	54	60	66	72
7	7	14	21	28	35	42	49	56	63	70	77	84
8	8	16	24	32	40	48	56	64	72	80	88	96
9	9	18	27	36	45	54	63	72	81	90	99	108
10	10	20	30	40	50	60	70	80	90	100	110	120
11	11	22	33	44	55	66	77	88	99	110	121	132
12	12	24	36	48	60	72	84	96	108	120	132	144

Thinking Games from around the World

Dou Shou Qi "The Jungle Game"
Ancient Chinese Game

Lion			Den			Tiger
	Dog				Cat	
Rat		Leopard		Wolf		Elephant
	≋ ≋	≋ ≋		≋ ≋	≋ ≋	
	≋ ≋	≋ ≋		≋ ≋	≋ ≋	
	≋ ≋	≋ ≋		≋ ≋	≋ ≋	
Elephant		Wolf		Leopard		Rat
	Cat				Dog	
Tiger			Den			Lion

Dou Shou Qi "The Jungle Game"

Objective: To capture all of the opponent's pieces **or** move one of your pieces into the opponent's den.

Playing Pieces/Set Up: Each player has eight animals; these may be cut out from the bottom of this page. (One set will need to be colored with one background color, and one with another…Or you may just want to use only one set of the animal pictures and use the set with words for the second set.) Each player puts their pieces on their own side of the board, on the labeled squares.

Playing: Players alternate turns. A turn consists of moving one of your own pieces one square horizontally or vertically (not diagonally). You may only land on a square occupied by an opponent's piece if their piece has the same number or a lower number than the piece you are moving. In that case you have captured the opponent's piece and it is removed from the board.

Special playing limitations:
Only rats can capture elephants and elephants may not capture rats.
Only the rats can move into the water spaces (but they can't capture elephants from the water).
The lions and tigers may jump across the water from one side to the other side as long as the space across from them is open or contains an opponent's piece they can capture. (But only if there's no rat in the way in the water.)
Animals may go into the trap spaces on their own side with no penalty.
Going into a trap on the opponent's side reduces an animal's number to zero (it can now be captured by anyone).

Game Over: The game is over when one player has removed all the other players' pieces by capturing them or has moved any one of their own pieces into the opponent's den.

Playing Pieces Could Look Like This:

8	7	6	5	4	3	2	1
8	7	6	5	4	3	2	1
Elephant 8	Lion 7	Tiger 6	Leopard 5	Dog 4	Wolf 3	Cat 2	Rat 1

Hints:
We printed the playing pieces and the board on cardstock. I made the game for a class, so I made an entire page of game pieces. I've included that page just in case you need it to. Just don't copy the next page if you don't need it.

8	7	6	5	4	3	2	1
Elephant 8	Lion 7	Tiger 6	Leopard 5	Dog 4	Wolf 3	Cat 2	Rat 1
8	7	6	5	4	3	2	1
Elephant 8	Lion 7	Tiger 6	Leopard 5	Dog 4	Wolf 3	Cat 2	Rat 1
8	7	6	5	4	3	2	1
Elephant 8	Lion 7	Tiger 6	Leopard 5	Dog 4	Wolf 3	Cat 2	Rat 1
8	7	6	5	4	3	2	1
Elephant 8	Lion 7	Tiger 6	Leopard 5	Dog 4	Wolf 3	Cat 2	Rat 1
8	7	6	5	4	3	2	1
Elephant 8	Lion 7	Tiger 6	Leopard 5	Dog 4	Wolf 3	Cat 2	Rat 1
8	7	6	5	4	3	2	1
Elephant 8	Lion 7	Tiger 6	Leopard 5	Dog 4	Wolf 3	Cat 2	Rat 1

Senet

Like many ancient games, the boards and pieces have been found,
but the rules must be figured out, so there tend to be many modern versions.
After much research, and trial and error, this is what we came up with.
You may find some other variations you enjoy.

History of Senet: Dozens of Senet games have been found in tombs in Egypt, dating back thousands of years. The thirty squares were believed to have originally been associated with the Egyptian calendar. It is considered the first known board game, and may have been a precursor to the modern game of Backgammon.

Objective: To be the first one to remove all one's pieces from the board, from the last square (bottom right).

Playing Pieces/Set Up: You will need one six-sided die and five playing pieces for each player.. The ten pieces are placed in alternating squares across the top row.

Blockades: Once two or more pieces of the same color are next to each other in the same row, they have formed a blockade. A blockade can be jumped over, but none of the pieces forming the blockade can be captured.

Playing: Players alternate turns. A turn consists of rolling a six-sided die, and then moving one of your pieces exactly that amount of spaces. Pieces move in the directions of the arrows, until they are moved off the board from the square in the lower right corner. You may not land on your own pieces. You may land on an opponent's piece as long as it is not part of a blockade. When you capture an opponent's piece, you move it to where your piece came from, and put your piece where their piece was.

Play Time: Our games took approximately 20 minutes each.

Senet

Game from Ancient Egypt

									Roll Exactly 1 to Remove
									Roll Exactly 2 to Remove
									Roll Exactly 3 to Remove Piece
									Return to Middle Spot
	Lose a Turn	"Middle" Spot							Mandatory Stop: Exact Roll Needed to Land Here

Brought to you by Creative Learning Connection
Copyright by Catherine Jaime, 2012

Latrunculi

This is a great two-player game for kids from eight to ninety-eight!

Boards

When we first encountered this game in a book of games it was with a simple four by four board. But when we started researching the game we found that most boards were believed to have been either eight by eight or eight by twelve. We had found the four by four game to be too simple, so liked the idea of playing on a bigger board. After trying almost a dozen different variations on the rules, we settled on two versions we liked the best – one on the eight by eight board and one on the eight by twelve board. The rules for both follow, with the simplest version first.

8 x 8 Board

Objective: To capture all of the opponent's pieces.

Playing Pieces/Set Up: Eight pieces of one color and eight of another (buttons, coins, stones, will work, too)

Each player puts their eight pieces anywhere on the board, one piece at a time. (Player A places a piece anywhere on the board that they want. Player B places a piece anywhere. And so it goes until all pieces are played.) No capturing takes place during this stage.

Playing: Players alternate turns. A turn consists of moving one piece to any one square in front, behind, or next to it (in other words, not diagonally).

Capturing: To capture a piece a player has to already have a piece on one side of the opponent's piece, and then in that turn place a piece on the opposite side. The captured piece is removed from the board and the capturer receives another turn.

Game Over: The game is over when one player has removed all the other players' pieces by capturing them. (The game would be a stalemate if both players got down to one piece left.)

8 x 12 Board

Objective: To immobilize your opponent's king.

Playing Pieces/Set Up: Each player has twelve regular pieces (pawns) and one unique piece (the king). The pawns are placed across the board in front of each of player. Each king goes on the second row, one square to their right of center.

P	P	P	P	P	P	P	P	P	P	P	P
					K						
						K					
P	P	P	P	P	P	P	P	P	P	P	P

Playing: Players alternate turns. A turn consists of moving a pawn as far as you want to go in a straight row in front, behind, to the left, or to the right of it (in other words, not diagonally) or moving the king one space in one of those directions.

Capturing: When a piece has an opposing piece to its right and left, or in front of it and behind it, it is captured and removed from the board. The capturer receives another turn.

Game Over: The game is over when one player has immobilized the other king. A king is immobilized if it cannot move in any of its four usual directions.

Play Time: Most of our games took 10 - 15 minutes each.

Origins

The game appears to have come forward in one or more of these forms from the ancient Roman era, though it was probably a variant of earlier similar Greek games.

Latrunculi (8x8 Board)

Game from Ancient Rome

Latrunculi (8x12 Board)

Game from Ancient Rome

Reversi

Reversi is played by two players on an eight by eight grid. The sixty-four playing pieces are reversible, dark on one side and light on the other. (An alternative would be to use coins – and have one player play with heads up and the other play with tails up.)

History

Reversi was originally developed in Victorian England in the late nineteenth century. The rules have only been modified slightly since then.

Objective

The objective is to have more pieces turned over to your color (dark or light) at the end of the game.

Starting the Game

The first four moves have to be on the start squares. Dark plays first, placing a piece on one of those four squares, followed by light on one of the remaining three, then dark, and finally light. After the first four pieces are placed, play continues, alternating between players. (See below)

Playing

After the first four pieces, a piece has to be placed in such a way that it sandwiches pieces belonging to the other player between the played piece and another of that player's pieces. When that happens, all the captured pieces are flipped. (See below)

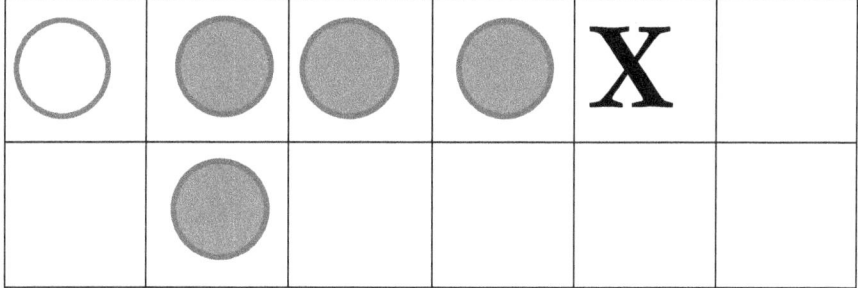

Placing a light piece where the X is will cause all the dark pieces that are captured between them to be flipped, becoming:

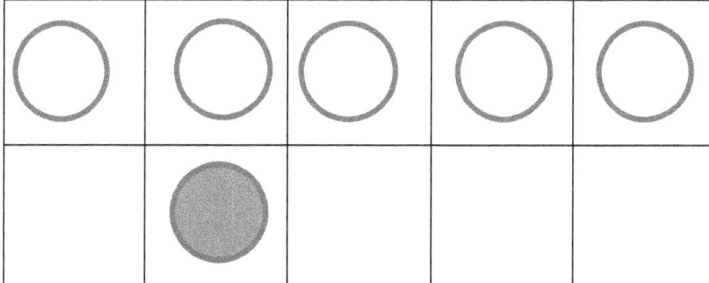

Game Over

Game continues until all sixty-four pieces have been played or until neither player has a legitimate play.

Reversi

Board Game from Victorian England

			Start	Start			
			Start	Start			

Reversi Playing Pieces

(Print page on cardstock, cut out pieces, one player can use the striped side, and one can use the plain side.)

Nine Men's Morris

(Game from the Colonial Period)

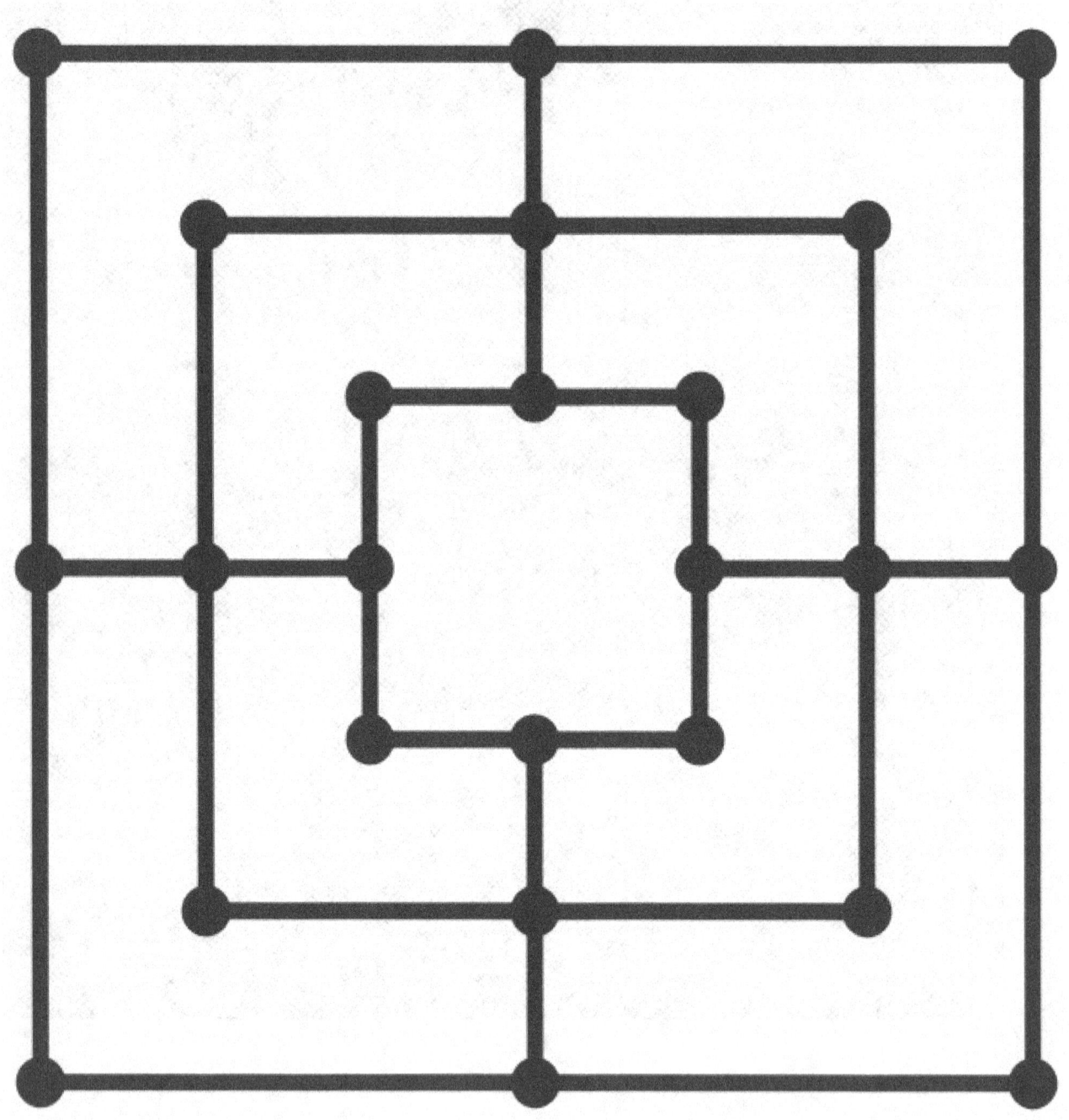

Nine Mens Morris

This is a great two-player game for kids from six to ninety-six!

Objective:
Each player starts with nine pieces (you can use pennies and dimes, two different color beans, or make small markers). The objective is to remove all of your opponent's pieces, or block them so they have no remaining moves.

Playing:
Pieces will be played on the small circles and can only be moved from one circle to an adjacent circle that is connected by a line: The game starts by players taking turns placing their pieces, one at a time, alternating turns. If a player gets three of their own pieces in a row, it is called a mill, and they get to remove one of the other player's pieces. The removed pieces are out of play for the remainder of that game.

After all nine pieces have been played on the board, the players take turns moving one of their own pieces from any circle to an adjacent circle. Again, any time a player has three connected pieces in a line, they have a mill.

Game Over:
Again, the game is over once either player has no more pieces or cannot make a legal move.

Play Time:
The game goes quickly, from set-up to completion. My kids would play several games in a row. The more they played, the better they got at determining the logic behind making their moves.

Origins:
This game has been played in various parts of the globe in various formats for at least centuries. If you look on line you will find a number of other variations on it, including Twelve Mens Morris.

www.ingramcontent.com/pod-product-compliance
Lightning Source LLC
Chambersburg PA
CBHW081240170526
45165CB00009B/3130